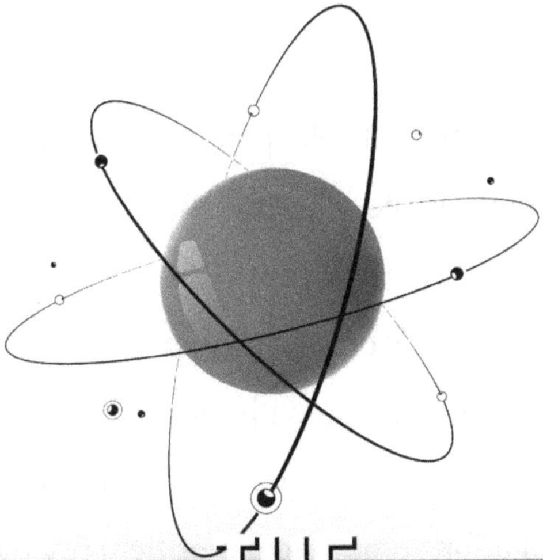

THE QUANTUM AGE

EXPLORING THE FRONTIERS OF REALITY AND POSSIBILITY

WRITTEN BY
PATRICIA DELACRUZ

"The Quantum Age: Exploring the Frontiers of Reality and Possibility"

COPYRIGHT

All rights reserved. No part of this publication may be reproduced, distributed, or transmitted in any form or by any means, including photocopying, recording, or other electronic or mechanical methods, without the prior written permission of the publisher, except in the case of brief quotations embodied in critical reviews and certain other noncommercial uses permitted by copyright law.

Copyright © Patricia Delacruz, 2024.

"The Quantum Age: Exploring the Frontiers of Reality and Possibility"

TABLE OF CONTENTS

Down the Rabbit Hole: A Quantum Adventure Awaits

Decoding the Quantum: A Dive into the Bizarre and Brilliant

Cracking the Quantum Code: Math Unveils the Microscopic

Warp Speed Computation: Quantum Computing to the Rescue!

Quantum Tech: A Leap into the Future's Playground

Down the Quantum Rabbit Hole: Questioning Reality One Paradox at a Time

The Quantum Frontier: Charting a Course to a Quantum Revolution

Unveiling the Quantum: A Universe of Endless Potential

Down the Rabbit Hole: A Quantum Adventure Awaits

Fasten your seatbelts, Alice! We're about to tumble down a rabbit hole, not into Wonderland, but into the mind-bending realm of quantum physics. Here, the familiar laws of our everyday world dissolve, replaced by a reality stranger than science fiction.

Imagine a universe where coins can land on heads and tails simultaneously, where a single particle can be in two places at once, and where the act of simply looking at something can change its entire existence. This isn't a dream — it's the bizarre and beautiful world of quantum mechanics.

"The Quantum Age: Exploring the Frontiers of Reality and Possibility"

Get ready to have your brain cells twisted! We'll delve into the mind-blowing concept of superposition, where particles defy logic by existing in multiple states at the same time. It's like having a cat that's both alive and dead... until you open the box, that is. This revolutionary idea, cooked up by the brilliant Erwin Schrödinger, shattered our understanding of reality and paved the way for the bizarre world of quantum theory.

Quantum physics is a historical treasure trove brimming with mind-blowing discoveries. From Max Planck's first glimpse into the quantum world in 1900 to Niels Bohr's groundbreaking work on the atom and Werner Heisenberg's uncertainty principle, these pioneers dared to question everything we thought we knew.

Prepare to witness the double-slit experiment, a legendary trick that proves particles can act like both waves and particles, depending on whether anyone's watching. It's like a mischievous electron doing the wave all by itself, then switching to a sneaky solo act when a scientist peeks in. Talk about a reality show with serious implications!

The ripples of quantum physics have splashed across science and technology, sparking revolutions that continue to this day. From the tiny transistors in your phone to the lasers that light up your concert, quantum mechanics is everywhere. Now, a new frontier beckons: quantum computers that promise to solve problems in minutes that would take regular computers billions of years.

So, grab your curiosity and a thirst for adventure. We're about to embark on a mind-bending exploration of the quantum realm, where the impossible becomes reality, and the universe's deepest secrets are waiting to be unraveled. This, my friend, is the quantum revolution – a journey to the heart of the unknown where the only limit is your imagination. Let's hop in and see where the quantum rabbit hole takes us!

Decoding the Quantum: A Dive into the Bizarre and Brilliant

Imagine a world where a coin can be heads and tails at the same time, where a single particle can be in two places at once, and where the act of looking alters reality itself. This isn't science fiction – it's the captivating realm of quantum physics, a revolution that's rewriting the rules of the universe.

Get ready to crack the code! We'll unravel the mind-bending concept of wave-particle duality, where particles defy logic by blurring the line between being a wave and a tiny marble. It's like a wave that can also whack a detector, a reality so strange it had even Einstein scratching his head.

But the weirdness doesn't stop there. Prepare to explore the uncertainty principle, a law that throws a wrench into our ability to predict the future. The more we try to pinpoint a particle's exact location, the fuzzier its momentum becomes, and vice versa. It's like trying to track a mischievous gremlin – the closer you get, the more it slips away.

The quantum revolution isn't just about mind-boggling theories. One of its most bizarre phenomena – quantum entanglement – is like having spooky twins. These particles become linked, sharing the same fate no matter how far apart they are. Measure one, and instantly the other knows, even across galaxies! Spooky, right?

The implications of this entanglement are mind-blowing. Imagine unbreakable codes based on this spooky connection, teleporting information across vast distances, or building super-powerful computers that solve problems in seconds that would take regular computers billions of years. The future is quantum!

So, buckle up, Alice! We're diving down the rabbit hole of the subatomic world, where the bizarre becomes brilliant and the impossible becomes reality. This is the quantum revolution, a journey where the only limit is your imagination. Let's see where quantum weirdness takes us!

Cracking the Quantum Code: Math Unveils the Microscopic

Imagine a universe written in code, a secret language that unlocks the behaviors of the tiniest particles. This code is the language of quantum mechanics, and within its equations lies the key to understanding the bizarre and beautiful subatomic realm. Join us on a journey to decipher this code, where math acts as our Rosetta Stone to the quantum world.

At the heart of this code lies the wave function, a mathematical equation that describes the probability of finding a particle in a specific location or state. Think of it as a map of possibilities, not a guarantee of where the particle definitively is. The ever-important Schrödinger equation governs how these wave functions evolve over time, reflecting the dynamic nature of the quantum world.

But the code goes beyond probabilities. Operators, like mathematical tools, represent physical properties like position or momentum. By applying these operators to wave functions, we extract information about the particle's behavior. It's like using the code to predict the outcome of a measurement – a powerful tool for understanding the quantum unseen.

Let's use the quantum harmonic oscillator as our practice problem. Imagine a ball trapped in a valley, constantly bouncing back and forth. Using the quantum code, we can derive its energy levels, revealing that the ball can only vibrate at specific frequencies – a signature feature of the quantum world, where things come in chunks, not smooth progressions.

Now, buckle up for a thought experiment: Schrödinger's cat. This famous paradox depicts a cat in a box, maybe alive, maybe dead, until we open the box and force it to choose a state. This spooky superposition, where a particle can exist in multiple states at once, is a core principle of quantum mechanics and a head-scratcher for even the best minds!

As we delve deeper into this code, we'll uncover the elegance and power of the math that shapes the subatomic world. Through real-world examples and mind-bending thought experiments, we'll unravel the mysteries of the quantum realm and gain a profound appreciation for the language that illuminates the universe's tiniest building blocks. Welcome to the captivating world of quantum mechanics, where math is our guide to the extraordinary!

"The Quantum Age: Exploring the Frontiers of Reality and Possibility"

Warp Speed Computation: Quantum Computing to the Rescue!

Imagine a computer that solves problems in minutes that would take regular computers billions of years. This isn't science fiction – it's the mind-blowing world of quantum computing, where the rules are rewritten and the power is exponential. Get ready for a wild ride into the future of computation!

At the core of this revolution are qubits, the superheroes of quantum computing. Unlike your regular computer bits (stuck at either 0 or 1), qubits can be both 0 and 1 at the same time, thanks to something called superposition. Think of it as a coin spinning in the air – it's both heads and tails until it lands. This lets quantum computers explore many solutions simultaneously, giving them a massive speed advantage.

But qubits aren't loners. They use special quantum gates to perform operations and talk to each other through a spooky phenomenon called entanglement. Imagine twins who know what the other is feeling, no matter the distance. Entanglement allows qubits to work together, making complex calculations a breeze.

This tag-team approach is what gives quantum computers their edge. With quantum algorithms designed to leverage superposition and entanglement, they can tackle problems that would bring regular computers to their knees.

Imagine cracking complex codes, optimizing logistics on a global scale, or simulating life-saving drugs – all in record time!

Speaking of codes, quantum computing might force us to invent entirely new ones! Some encryption methods used today will become vulnerable to the mighty qubit. But fear not, advancements in quantum cryptography are on the horizon, using the same principles to create unbreakable codes.

The possibilities are endless. Quantum computers can revolutionize materials science, design unbreakable communication networks, and even help us understand the universe better. It's like giving scientists a superpowered microscope to see the world in ways we never imagined.

So, buckle up and get ready to warp into the future of computing! Quantum computing is here to break the boundaries of what's possible, ushering in a new era of innovation and discovery. The only limit is our imagination!

"The Quantum Age: Exploring the Frontiers of Reality and Possibility"

Quantum Tech: A Leap into the Future's Playground

Imagine a world where medical diagnoses happen with mind-blowing precision, where communication is uncrackable, and where entire networks hum with the power of the quantum realm. This isn't some distant dream – it's the exciting world of quantum technologies, where innovation is redefining the future.

Get ready to explore the cutting edge! First up, quantum sensors. Think of them as super-powered detectors. Using the weirdness of quantum mechanics, they can see things with mind-boggling detail. Imagine doctors using these sensors to peer inside your body, spotting diseases at their earliest stages, or engineers using them to uncover hidden resources deep underground.

But that's not all. Quantum tech is also revolutionizing how we measure things. Quantum clocks keep time with incredible accuracy, making them perfect for tasks like ultra-secure financial transactions or guiding self-driving cars. And quantum magnetometers can map the Earth's magnetic field with incredible detail, helping us understand everything from weather patterns to volcanic activity.

Now, let's talk about communication. Quantum communication uses the spooky phenomenon of entanglement to create unbreakable codes. Imagine information becoming like a pair of linked coins – changing together no matter the distance. This means hackers wouldn't stand a chance! Quantum communication promises ultra-secure networks for everything from banking to national security.

But wait, there's more! These secure networks can be linked together to form a "quantum internet." Think of it as a superhighway for information, carrying massive amounts of data across vast distances with incredible speed and security. This could power the next generation of the internet, connecting quantum computers across the globe and unlocking unimaginable possibilities.

The potential of quantum technologies is truly mind-blowing. From revolutionizing healthcare and communication to giving us a deeper understanding of our planet, these technologies are poised to change the world as we know it. So, join us on this exciting journey into the future, where the playground of quantum mechanics becomes the launchpad for a brighter tomorrow!

Down the Quantum Rabbit Hole: Questioning Reality One Paradox at a Time

Fasten your seatbelts, Alice! We're tumbling down a rabbit hole, not into Wonderland, but into the mind-bending world of quantum reality. Here, the familiar laws of physics dissolve, replaced by a reality stranger than science fiction. Get ready to have your reality questioned and your mind boggled!

At the heart of this quantum weirdness lies the wave function, a mathematical equation that describes the probabilities of finding a particle in a particular state. Imagine a coin that's both heads and tails at the same time, but only reveals one side when you look. This wave function only solidifies into a concrete reality – a particle here, a spin there – when we make a measurement. This strange behavior is the foundation of the Copenhagen Interpretation, the reigning theory of quantum mechanics. It suggests that observing something forces it to choose a state, collapsing the wave function and shaping reality itself. Pretty trippy, huh?

But hold on, the plot thickens! The Copenhagen Interpretation isn't the only game in town. There are many ways to interpret the bizarre world of quantum mechanics. The Many-Worlds Interpretation, for example, suggests every possible outcome of a measurement spawns a new universe. So, if you look at that quantum coin, one universe sees heads, another sees tails, and countless others exist in a superposition of both. Mind blown? You bet!

Pilot-wave theory throws another wrench into the works. It proposes hidden variables, like a secret script, dictate how particles behave. This removes the randomness of the Copenhagen Interpretation, but raises the question: who wrote the script, and where is it hidden?

As we delve deeper, we confront profound philosophical questions. Does consciousness play a role in shaping reality? Does the act of observing something actually create it? Pioneering physicist John Wheeler said it best: "No phenomenon is a real phenomenon until it is an observed phenomenon." Spooky, right? Experiments by Anton Zeilinger hint that observation can influence particles across vast distances, blurring the lines between observer and observed.

The more we explore the quantum realm, the more we realize the answer might be "we don't know yet." Quantum mechanics challenges our assumptions about reality, existence, and consciousness. Embrace the ambiguity! This is a journey of discovery, not a destination with all the answers. So, join us as we delve into the mysteries of the subatomic universe, question reality one paradox at a time, and embark on a quest for understanding that pushes the boundaries of human knowledge. Welcome to quantum reality – a rabbit hole filled with wonder and endless possibilities!

The Quantum Frontier: Charting a Course to a Quantum Revolution

Imagine a future where computers solve problems in minutes that would take regular machines billions of years, where communication is uncrackable, and sensors can peer into the world with unimaginable detail. This isn't science fiction – it's the uncharted territory of the quantum frontier, brimming with challenges but overflowing with potential. Join us on a voyage of discovery as we navigate the hurdles and seize the opportunities that lie ahead!

The road to a quantum revolution isn't paved with magic pixie dust. There are roadblocks to overcome. Scaling up these technologies is a major hurdle. Right now, quantum systems are like tiny, prima donna computers – they work best when small and babying them. The challenge is making them bigger and more robust, able to handle complex tasks without throwing a tantrum.

Another roadblock is keeping these quantum systems happy. They're very sensitive to their environment, like fussy toddlers. Any outside noise or temperature change can disrupt their delicate calculations. The key is figuring out how to extend their attention span, or coherence time, so they can focus on the job at hand.

Finally, there's the constant chatter that disrupts the quantum signal. Imagine trying to do math in a noisy classroom. That's the world of quantum tech – lots of background noise messing with the calculations. We need to find ways to quiet things down and make the signal clear.

But fear not, intrepid explorer! These challenges are not insurmountable. Breakthroughs in materials science and engineering are promising to build bigger and better quantum systems. Just like engineers learned to build more powerful computers, we'll learn to tame the quantum beast. New techniques to correct errors and innovative materials that shield out noise are on the horizon.

The payoff for overcoming these challenges is immense. Imagine a future where quantum computers tackle problems in drug discovery or materials science that would take regular computers longer than the universe has existed. Unbreakable communication networks could safeguard sensitive information, and super-sensitive sensors could revolutionize medicine and environmental monitoring.

The future of quantum is not just about individual technologies – it's about a powerful synergy. Quantum computers, communication, and sensors working together could usher in a new era of scientific discovery and technological advancement. Imagine a world where quantum machine learning cracks complex scientific problems or quantum networks connect scientists across the globe in real-time.

The quantum frontier is a vast landscape, brimming with challenges and opportunities. By embracing these hurdles as stepping stones, we can unlock the transformative potential of quantum technologies and propel ourselves towards a brighter future. So, buckle up, fellow explorers! The journey to the quantum revolution begins now!

Unveiling the Quantum: A Universe of Endless Potential

We've reached the end of our whirlwind tour of the quantum realm, but the journey of discovery is just beginning. Let's take a moment to recap the mind-blowing world we've explored and ponder the exciting future that awaits.

We've grappled with the bizarre principles of quantum mechanics – superposition, where particles can be in multiple states at once, and entanglement, where particles become spookily linked across vast distances. We've seen how these principles hold the key to unlocking revolutionary technologies like quantum computers and unbreakable communication networks.

Sure, there are challenges. Scaling these technologies up and keeping them functioning smoothly requires innovation. But with breakthroughs in materials science and error correction on the horizon, these hurdles seem less daunting.

The potential rewards are simply too immense to ignore. Imagine computers that crack problems in seconds that would take regular machines eons, or sensors so sensitive they can peer deep into the mysteries of the human body. The possibilities are endless, promising solutions to problems in healthcare, materials science, and beyond.

The beauty of the quantum revolution is that it's not just about individual technologies. Imagine a world where quantum computers, communication, and sensors work together in a powerful symphony. This synergy could usher in a golden age of scientific discovery and technological advancement.

But the quantum revolution goes beyond the practical. It challenges our very understanding of reality. The observer and the observed become intertwined, forcing us to question the nature of consciousness and existence itself.

As we stand on the precipice of the quantum age, let's embrace the unknown with curiosity and a sense of wonder. This isn't just a scientific or technological shift; it's a philosophical one. By approaching the quantum revolution with open minds and a thirst for knowledge, we can push the boundaries of human understanding and illuminate the mysteries of the universe.

So, dear reader, don't be a stranger! Join us on this ongoing exploration of the quantum realm. Together, let's unlock the secrets of the universe and shape a future brimming with possibility. The quantum age is here, and the future is ours to create!

www.ingramcontent.com/pod-product-compliance
Lightning Source LLC
Chambersburg PA
CBHW050253230526
45470CB00005B/2239